# How to make mind reader

I0504242

- Read the mind

- Induct thought

- Induct dream

- Evoke spirit

  By mind reader

## MOHAMMAD HOSSEIN

## SHAFIEI

# CONTENTS

# Identify concepts

When I was a child I had many questions about myself and my surroundings world, questions such as how I dream when I am sleeping, how some of the dreams come true, how we are aware about something that might happen in the future, why we dream, how we can understand what is happening in a dream is not a reality, why we imagine it's reality when we are dreaming while we are sleeping, how can we realize that we are dreaming and many issues that perhaps was routine for other children was strange in my mind.

Fifteen years ago when I thought about the human existence reasons and possible ways to communicate between people, I realized that we have ability to communicate without speech and through the brain device. I justify that every word has a concept that before the expression it is made in the brain and then pronounced with sounds and words in a language. Therefore, a concept is created with another nature before voice expression.

If we have the ability to identify and reproduce this nature, we can translate it and then no matter that the person speaks in what language because the new known nature indicates the concept and not the word. This new nature is brain waves. Thought is the wave and can be released in space by a device called brain. Each part of the emitted waves of the brain contains a concept.

In fact, every brain wave is like a letter that makes a word when adds to other letters. If we show an image to a person, the brain releases a wave upon seeing the image, similarly about words or letters. That means every letter, every word, and every image, and in general everything that is conceptual and the person has the ability to sense and comprehending it, has its own wave in the brain.

for example:

 = A

By recording each one of these waves, it's possible to read the meaning of each wave and ultimately the whole meaning of the brain waves in the person's mind. You can actually have a database similar to the Wave Dictionary that translates each wave. Also, by making similar waves by the wave generator or oscillator, it's possible to instill a mental image to the person. In fact, it is possible both to read and to induce thought by the device.

The data is stored on the brain device where it can be accessed when needed and compared or processed if needed. Each sensation and mind's image causes each brain cell to be in a state that emits a specific signal. The sum of these signals is a sign of sense and image. By stimulating the relevant parts of the brain can call and download person's thoughts and memories, or by sending waves can induce or upload a thought.

Also, if the brain's neurons be enough sensitive to activate and respond to weak external frequencies and the person's brain system be capable to receiving the same frequency and similar waves of another person's thought, they will be able to communicate mentally with each other.

The best use of this issue in that time that came to my mind was about people who are not able to speak. Also uploading and inducting thought to people who lost their memory so they could communicate with others using this device. And then about animals that through it effectively communicate with them was possible.

# Database

# For Identifying Brain Waves

By measuring the characteristics of the wave propagated from the brain when one thinks about a particular concept and the image of that concept is made in his or her mind, one can record that wave as a reference data or an identifying wave. By recording different waves of mental images of different concepts, a brain wave identification database can be created.

# Brain

The human brain is a wave's receiver, registrar, analyzer and transmitter. Thoughts are composed of a series of waves that each wave marks the smallest component of thought, such as letters in a word. Including having the characteristics of each wave it's possible to analyze and translate the complex waves. By this way all your thoughts are being revealed before you tell it, and other people can use this method to create the thoughts in your mind that you are reluctant to think about it.

A part of the brain's task is to receive and disseminate waves of thought. This means your thoughts have the ability to publish in the environment.

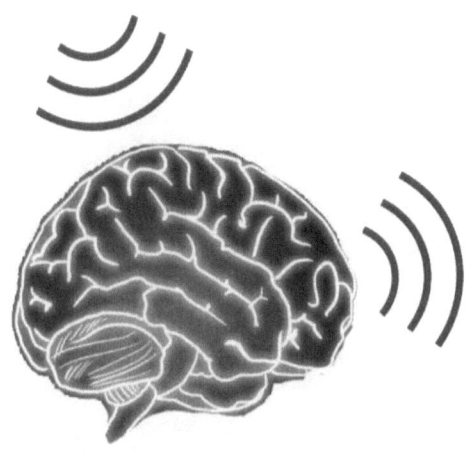

Image: Brain is a wave receiver and transmitter

Thinking means the production and propagation of waves in the environment, and these waves can be recorded by the wave's receiver device or thought reader. This means it can download and analyze your memory and your thoughts and reach to your memories.

To get the brain waves there is no need to connect the wires and electrodes to the head, but the waves are received from the distance via the wireless receiver.

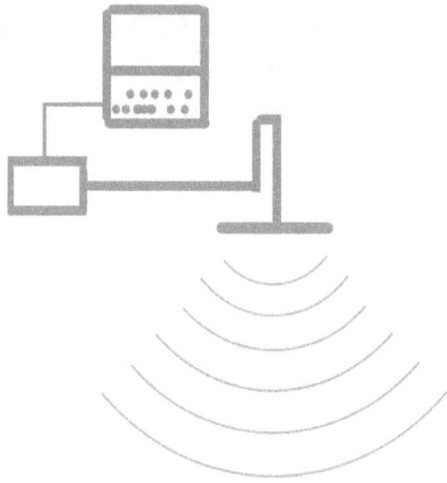

Image: Mind reader attract and transmit the waves

Every image in your brain is interpreted as electromagnetic signals, so it can create specific images in your mind by sending electromagnetic waves with a specific frequency and intensity from a great distance, such as through walls and ceilings to your home and your head. You will understand it as the thought or the dream.

The different parts of the brain and glands can be stimulated and activated by this way. By stimulating the brain and the nervous system it's possible to activate your memory and your memories and force you to meditate or browse your brain memory.

Information and concepts and images can be repeatedly transmitted to your brain through the wave's producer device and in so far as repeated that you have to think about it. It create an image in your mind without your will. After that created image in your brain memory will read out by stimulate your brain and you will think about it without your will.

This can be called a virus of mind, a thought or mental image of something that is imposed on you without your consent and that you have no control over.

In this way, by using the device of induction and transmitting electromagnetic waves over the head can generate good thoughts or evil thoughts in one's mind and even encourage one to decide and do something contrary to his or her will.

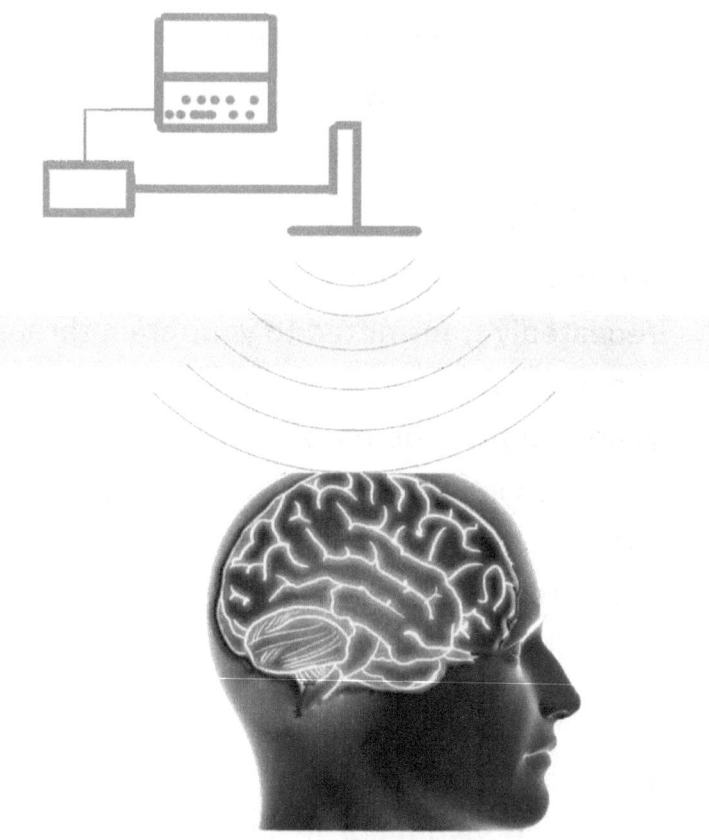

Image: Waves leads over the wall

An unwanted mind's image or viral image of the mind that can be active in your mind for a longer time can more effectively control you and affect your life because your thoughts actually shape your life.

Every action you take is the result of a decision that your thought is taken. These decisions is taken after receiving data and comparing the data with the mind reference information and reviewing the results. The decisions can be made very quickly or slow. In the alert and high irritation mode, quick decisions are taken and rapid response is made. Different methods and factors cause high irritability, which is not the subject of this discussion.

# Brain waves frequencies

The frequencies of the main brain waves are in the specific range of half to thirty Hz. Frequency of electrical pulses of nerve cells in the brain at any given moment is your mental image and thought at that moment.

The brain waves are named as follows by neuroscientists according to the frequency range:

Delta waves of half to four Hz

Theta waves of four to eight Hz

Alpha waves of eight to twelve Hz

Beta waves, Twelve to thirty Hz

Gamma waves thirty to one hundred Hz

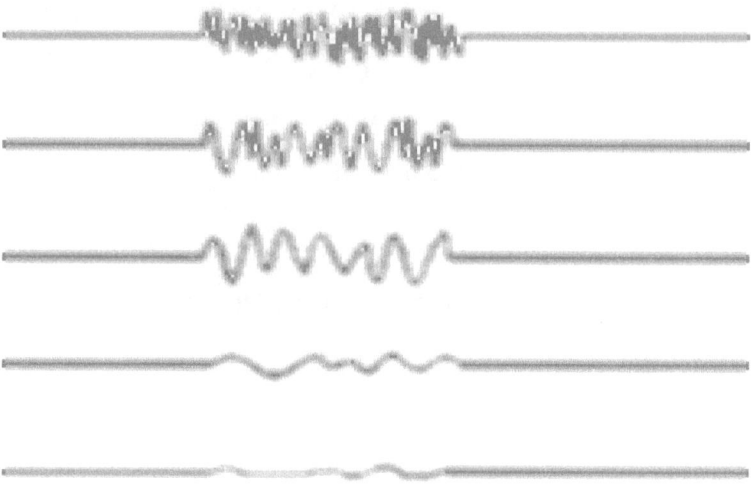

Image: The brain waves contain Beta, Alpha, Delta, theta, Gamma

In each one of these frequencies the brain activity varies, it is less active at low frequencies, and the higher brain's activity will cause the higher brain's frequency. In fact, more number of activated neurons will cause the greater wave and also the sum of the frequency of their activity will be more.

Cognition or understanding is reach to a rational outcome based on comparing observations and evaluations with the subjective scale, so if each of these, means observing that is input data and the mental scale that is the information stored in the person's mind as a reference for comparison, is changed so a different outcome and a different perception of the subject will be cause, that will ultimately lead to a different decision.

The response resultant that brain cells make to a subject is how we feel about it.

In order to make a device that be able to read the mind or to induce thought, first, we need to know how process is thinking, what performance makes thinking, how you think about something and basically why do you think.

The thought is wave, the specific waves that are produced, received, processed, stored and spread by the brain's device. Every wave is known with its own specifications like the frequency, range, intensity and how release.

# Thinking process

Thinking begins with a question. The question that arises in your mind can be the need or the difference between what you expect and what is happened or the change or the sense of duality and difference in the subject that drives you to think. Thinking begins with a topic. This topic can be a data or mental image of what you are thinking about. So the topic is part of the thinking.

Subjects are generally received by the senses, and the memory is able to store an image of the data that received by the senses and sends them to the brain's scale for comparative analysis. For example, when you watch a scene or you hear a word or you smell, you receive data about it through the eye or ear or nose, and then send it to the brain's comparator for analysis and identification after being converted to electrical pulses. Also you use the topics that have stored in your memory when you compare the information in your brain's memory to draw conclusion.

Comparing information is the main part of the thinking process. When you choose a topic for thinking, it compares to the rest of the information that was stored in your brain memory. You have created a scale in your mind that is perfectly correct of your own mind's viewpoint. This is actually the same things that you have experienced or learned to you, and you have saved them in your brain memory as a reference, and you compare anything with that to prove it is right or wrong.

Reason of thinking can be found in its result. You start thinking to getting a result about the issue that is in front of you. This result can reject or accept a subject and ultimately lead to a decision or action. Or create a new concept or remove or refine a concept. So a part of the thinking process is conclusion.

Any phenomenon which occurs change in that if it is known and defined for the two beings can have meaning and is data. Therefore, if a wave is defined for two devices, one receiver and one transmitter, it can transmit a concept called data and appear as information after identification, analysis, and translation.

The received data at a single moment are in one state only, they are true or false, and cannot be excluded of the two. One data at a time cannot be both true and false, if you conclude that the subject or data is also true and incorrect, know that the observations and preliminary data were incorrect or incomplete, wrongly analyzed or compared, or incorrect conclusions.

To transmit information, a transmitter, a receiver, a platform or channel of transmission between the receiver and the transmitter is required.

# Mind reader's components

The mind reader device is composed of an electrical power supply, a wave receiver, a wave simulator and a wave analyzer with the database, an oscillator or a wave maker, a wave's transmitter, an antenna and a monitor.

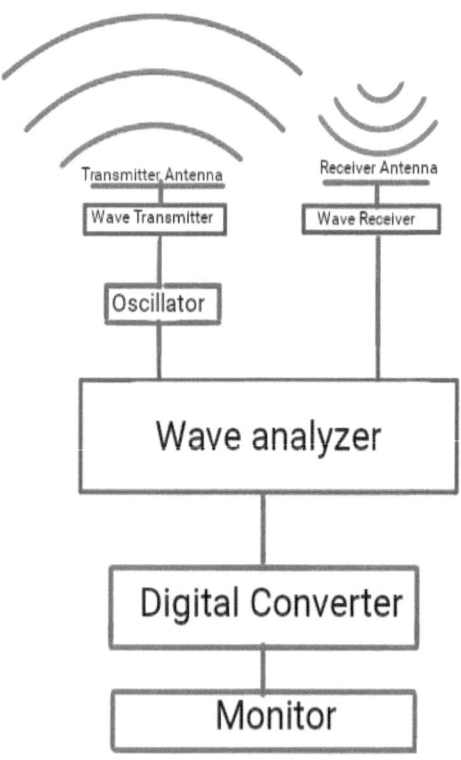

Image: mind reader diagram

## Receiver

The waves produced by the human brain are received by the receiver. The receiver receives the waves and also the information that is taken in the wave.

## Wave Analyzer

Comparing the received wave information with information of the defined wave reference and identifying the wave performed by analyzer. An encyclopedia or database of wave types specifications is defined as the reference and the incoming waves and the meaning of each one is recognized with this encyclopedia or identification database. This database can be defined as digital signals and the received waves are translated into digital signals before analysis and then into images or concepts.

Comparator

| References Waves |
| :---: |
| Received Waves |

Image: Comparator recognize received waves by comparing with reference waves

This encyclopedia can also translate the image into wave and be induced to mind by the inducer device. As previously explained, every concept in the mind that emerges as an image is a wave that created from the sum of the signals emitted by the brain's neurons. The sum of these images and the waves make the database or information source of comparator that is used by both the mind reader and the thought inducer.

## IMAGE CONVERTOR

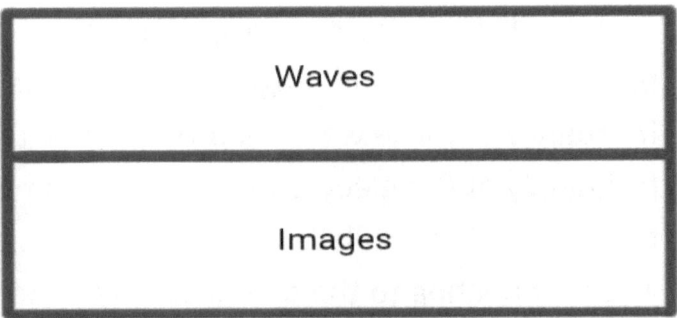

Image: Image convertor convert image to wave and wave to image by using references database

These concepts and images are translated in the determined frequency range of the waves and can be integrated into specific frequency and amplitude waves.

## Oscillator

Each oscillating electric charge in space produces electric and magnetic waves. The frequency of these waves is proportional to the frequency of the electrical charge generating the waves. The oscillator is actually an orbit that after connecting to the power supply, reaches stable oscillations.  At first range of oscillating is rose till comes to the specified range and there the unstable rise stops and then the oscillation continues at a certain amplitude. In fact, we produce the desired waves by oscillator.

## Transmitter

The produced waves are directed to the target by the transmitter so that the information contained in the wave is transmitted by the transmitter to the target.

When brain waves are receiving, any stimuli that cause to stimulate and respond nerve cells in the brain can be used to separate disturbing waves from brain waves, such as sound or light.

With this reaction, nerve cell waves of the emitted waves of devices and lifeless organisms are detected. The human nerve cells react to the shock or light sound, making a leap in the form of waves, if the devices and organisms that emit waves are not reactive towards these stimulants.

# Edit the Dreams

You can also induce images and scenes during sleep, with no wires and from behind the walls, which can be viewed as a dream. In this way, response of person to those scenes can be measured. Also asked the person some questions that he or she would respond to them and his or her desire for that image or concept can be determined. You can also create situations in a planned dream that person does not want to attend.

Dreaming is a process created by the alternating activity of some brain cells when some other cells are inactive, during that, the information is already stored in memory or received data in different ways at the same moment is reviewed, and some parts of them are combined or separated and a mental image is created. When one is sleeping, he / she supposes these mental images are real and imagines himself / herself observer of events and reacting to them.

Different stimuli can activate different brain's cells. One of these stimuli is electromagnetic waves including information that can directly transmit information.

# Evoke spirit

It is also possible to do that with a data transmission network such as the Internet from the other part of the world. So by thought's induction and readout equipment that emits waves over the person and with network equipment and data transmitter can inquire questions of his or her subconscious from the another points of the world. In fact, it is the summoning of the soul or conjure spirit that the subconscious of the person present and responding to others in the other side of the world without being aware that it is planned and directed.

The same method is used to communicate between two people in two different places. That is, two sleeping people question each other and examine reaction of each one of them.

According to the material, humans have the ability to read others' thoughts using the mind-reader apparatus, as well as the ability to change the thoughts and dreams of other person and also change decisions using the thought inducer.

In addition, one can remotely interrogate a person through his or her subconscious in sleep, and in addition to accessing his or her memory, can reach to his or her wishes and desires. Also by repeat a specific thought or mind's scene can forced him or her to do something.

In the dream, the person has no control over his or her thoughts and what is called the self-conscious does not exist and only the subconscious will respond to the needs and impact of the environment.

So at sleep the desires of a person to perform act can be gauged so that the person is unable to hide it, although the desire varies in different situations, and the desire at sleep is not a sign of action at waking time. In fact, this is the way to measure the scale of one's mind.

 The self-consciousness creates rules in the mind that will keep the person within a certain framework for action and encourages him or her to examine them before conversion to action and conform tendencies to its laws.

Each one of us may had make billions of thoughts in our minds in our lifetime. Each one of these thoughts is about a particular subject and tends towards specific direction. Many of them are aligned and tend to the same subject. This desire for a subject creates a flow that builds one of your personality dimensions, and you will have that aspect of your personality and desire with yourself for the rest of your life. But your overall personality is resultant of these tendencies and thoughts flows that have changed over time. Whatever you think more about a subject and learn more about it and master on it, the bold and more dominant's stream of your thought will tend to that point, and your personality will tend to that topic, and your character will be highlight by that specification. But desires can be changed by different ways. So the personality of person can be changed.

So by using mind reader device you can:

Read the other person's thought and see his or her mental images without he or she realize it, because it is done over the wall and by sending waves which passing through the walls.

Immerse the person in the thought you have planned without having the will

Change the mind's scale of person

Make the person to sleep

Make the person see the planned dreams and images that he or she may not want to see

Create different situations in the dream and observe the reaction of the person to that situation

Wherever you are, evoke spirit and summon person's subconscious when he or she is sleeping and ask questions as he or she answer your questions.

According to the contents we need to find a way to prevent abuse of the mind-reader devices and thought inductors for control and domination over people and society.

The End

"Philosophical question, Physical answer"

Other books that are available at amazon.com and other book stores:

21 points to become philosopher

Review a constitution

Email: mohammadhosseinshafiei95@gmail.com

www.ingramcontent.com/pod-product-compliance
Lightning Source LLC
Chambersburg PA
CBHW030541220526
45463CB00007B/2927